SCIENCE EXPERIMENTS

Air

Om Books International

Om Books International

Reprinted in 2025

Corporate & Editorial Office
A-12, Sector 64, Noida 201 301
Uttar Pradesh, India
Phone: +91 120 477 4100
Email: editorial@ombooks.com
Website: www.ombooksinternational.com

Sales Office
107, Ansari Road, Darya Ganj
New Delhi 110 002, India
Phone: +91 11 4000 9000
Email: sales@ombooks.com
Website: www.ombooks.com

ISBN: 97-893-52763-10-8

Printed in India

10 9 8 7 6 5 4 3

CONTENTS

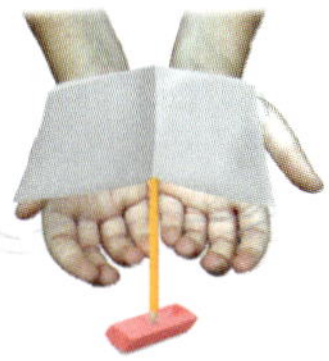

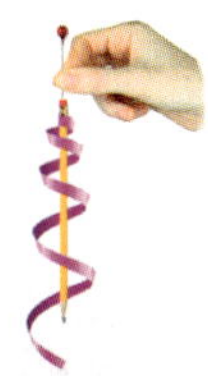

Duration of Experiment

This icon indicates approximately how long the experiment will take, including the set-up time.

Difficulty Level

The black bars in this icon indicate level of difficulty in doing this experiment with one icon depicting the least level of difficulty.

Adult Supervision

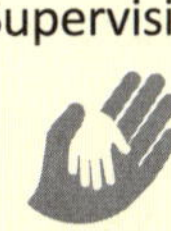

This icon indicates that parental supervision is required. Always have an adult beside you.

NOTE: All activities and experiments described in this book should be ideally performed with adult supervision. Parents and guardians are advised to oversee the children and assist them with any difficult or potentially harmful tools. Common sense and discretion are advised. The publisher does not assume any responsibility for any injuries or damages arising from any activities.

Tsp - teaspoon Tbsp - tablespoon ml - millilitre l - litre gm - gram 1 cup - 250 millilitre cm - centimetre

Bubble Within a Bubble

Duration of Experiment: 15 mins.

Difficulty Level:

What you need:

Plastic cup

Straw

Wire loop

Bubble mix

Water

What to do:

Step 1. In case the bubble mix isn't available, you can make your own mix by mixing water, washing soap and glycerine.

Step 2. Turn the cup upside down and place it on a flat surface like a table.

Step 3. Wet the surface of the cup with water.

Step 4. Use the wire loop to blow a bubble on the base of the cup, which is now on top.

Step 5. Wet the straw with the bubble liquid by dipping the straw in the liquid.

Step 6. Gently slide it into the bubble and blow a smaller bubble. You now have a bubble inside a bubble!

SCIENCE AROUND US

How do bubbles form?

Bubbles are pockets of soap and water that are filled with air. When soap and water are mixed together and air is blown into the mixture, the soap forms a thin skin or wall and traps the air, creating a bubble.

What just happened? Bubbles are air trapped within a liquid ball. Contact with a dry surface makes it burst. An object that is wet and soapy itself can easily enter the bubble without bursting it.

Incredible Hoop Glider

Duration of Experiment: 15 mins.

Difficulty Level:

What you need:

Straw

Tape

Scissors

Card paper

What to do:

Step 1. Cut the card paper into three strips (2.5 cm × 13 cm).

Step 2. Make one large loop out of two strips.

Step 3. Make a smaller loop out of the third strip.

Step 4. Tape both the loops to the ends of the straw.

Step 5. Hold your glider by the straw, angle it upwards and launch it into the air!

What just happened?

The two loops keep the straw balanced. The large loop creates a 'drag' which keeps it airborne, and the smaller one keeps it from turning off course.

Air Twirly

Duration of Experiment: 10 mins.

Difficulty Level:

What you need:

Eraser

Scissors

Pin

Pencil (with an eraser at the end)

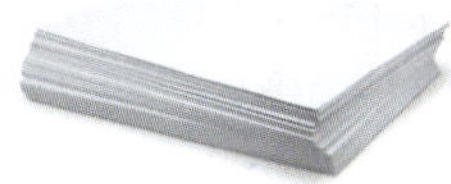

Paper

What to do:

Step 1. Cut a square out of the paper. Fold it diagonally.

Step 2. Stick the tip of the pencil in the eraser and stick the pin into the bottom of the pencil.

Step 3. Place the centre of the paper on the pin.

Step 4. Rub your hands together and place them below the paper. The paper starts spinning!

What just happened?

The warmth from your hands heated the air around it, which began to rise, causing the twirly to spin.

Eggs-periment

Duration of Experiment: 10 mins.

Difficulty Level:

What you need:

2 bowls

Boiled egg

Water

Plastic bottle

What to do:

Step 1. Put the bottle in the bowl of hot water for about 5 minutes.

Step 2. Move the bottle to the bowl of ice water. Wet the egg and place its pointed side down in the bottle opening. The egg gets sucked into the bottle!

What just happened?

When the air in the bottle is heated, the air pressure in the bottle increases. As it cools, the pressure decreases. The air pressure from outside pushes the egg into the bottle.

Magic Finger

Duration of Experiment: 10 mins.

Difficulty Level:

What you need:

Tin can

Bowl

Drill

Water

What to do:

Step 1. Use the drill to bore holes at the bottom of the can. Make one hole on the lid.

Step 2. Hold the can over the bowl and try to pour water in it. It drains out from the bottom.

Step 3. To stop the water from draining, you need to use your 'magic finger'.

Step 4. Pour water in the can, close the lid and quickly shut the hole in the lid, with your finger.

What just happened?

By closing the lid, you stopped air from entering the can and the air outside the can presses upwards on the underside of the can to prevent the water from running out.

Levitating Spiral

Duration of Experiment: 10 mins.

Difficulty Level:

What you need:

Paper

Pencil

Scissors

String

Candle

Don't hold the spiral too close to the flame. It could catch fire.

What to do:

Step 1. Draw a thin spiral on a piece of paper and cut it out.

Step 2. Light the candle.

Step 3. Tie the spiral to the pencil using the string. Hold this over the candle.

Step 4. Watch as the paper rises by itself.

What just happened?

The air around the candle gets heated. As the air gets warmer, it expands and becomes lighter. It rises. Since the spiral is light, it rises along with the air.

Inflating Balloons

Duration of Experiment: 10 mins.

Difficulty Level:

What you need:

Balloon

Soda bottle

What to do:

Step 1. Pour half the soda out.

Step 2. Put the balloon over the mouth of the soda bottle.

Step 3. Shake the bottle a little bit.

Step 4. Watch the balloon inflate!

What just happened?

The small bubbles that you see in soda are carbon dioxide bubbles. When you shake the bottle, the gas is released and it fills the balloon.

Bendy Air Trick

Duration of Experiment: 5 mins.

Difficulty Level:

What you need:

Paper

Tape

Book

Bottle

Water

Scissors

Table

What to do:

Step 1. Cut the paper into a strip (2 cm × 10 cm). Fold one end.

Step 2. Tape the paper to the table such that the longer portion of the strip is standing perpendicular to the table.

Step 3. Place the book between yourself and the paper. Try to make the paper flutter by blowing on the book. You won't be able to do so.

Step 4. Fill half the bottle with water and place it where the book was. The paper will flutter.

SCIENCE AROUND US

Aerodynamics

This property of air to 'bend' around curved objects is the reason why ships, cars and aeroplanes are the shape that they are. The streamlined shape of planes and ships allow them to move faster.

What just happened?

The book blocks the air that you are blowing. The air deflects back at right angles. But the air bends around the curved surface of the bottle, causing the paper to flutter.

Make a Hot Air Balloon

Duration of Experiment: 10 mins.

Difficulty Level:

What you need:

Duct tape

Large plastic bag

Hair dryer

What to do:

Step 1. Through the holes in the neck of the plastic bag seal it with duct tape to make the opening of the bag narrow.

Step 2. Blow hot air into the plastic bag using the hair dryer directed through the now small opening.

Step 3. After a few seconds, when the plastic bag is full of hot air, turn the hair dryer off and let go of the bag.

Step 4. The bag begins to rise towards the ceiling and hovers for some time.

What just happened? Hot air expands and rises, carrying your 'hot air balloon' along with it.

Snake Spiral

Duration of Experiment: 10 mins.

Difficulty Level:

What you need:

Scissors

Pencil (with an eraser at the end)

Pin

Paper

What to do:

Step 1. Draw a spiral on the paper and cut it out.

Step 2. Loosely pin one end of the spiral to the rubber on top of the eraser.

Step 3. Hold the pencil over a stove.

Step 4. Watch the spiral begin to twirl!

What just happened? The hot air above the stove rises, causing the spiral to rise along with it.

Self-inflating Balloon

Duration of Experiment: 20 mins.

Difficulty Level:

What you need:

Balloon

Pan

Empty bottle

Water

What to do:

Step 1. Stretch the balloon over the mouth of the empty bottle.

Step 2. Put the bottle in the pan of hot water for a few minutes.

Step 3. You should be able to see the balloon inflating soon.

What just happened?

As the air inside the bottle starts getting heated, it begins to expand. This air is trapped by the balloon, which begins stretching.

Blowing Balloons

Duration of Experiment: 20 mins.

Difficulty Level:

What you need:

Lemon

Straw

Balloon

1 tsp baking soda

Plastic bottle

Water

What to do:

Step 1. Pour the water into the plastic bottle.

Step 2. Add the teaspoon of baking soda and stir it around with the straw until it has dissolved.

Step 3. Squeeze a few drops of lemon in and quickly put the balloon over the mouth of the bottle.

Step 4. Watch as your balloon inflates.

What just happened?

As the air inside the bottle starts getting heated, it begins to expand. This air is trapped in the balloon, which begins inflating.

Balloon Expansion

Duration of Experiment: 20 mins.

Difficulty Level:

What you need:

Balloon

Yeast

2 tsp sugar

Plastic bottle

Water

What to do:

Step 1. Fill the bottle with an inch of warm water.
Step 2. Add yeast and gently swirl the bottle a few seconds.
Step 3. Add the sugar and swirl it around some more.
Step 4. Put the balloon over the neck of the bottle.
Set it aside in a warm place for 20 minutes.
Step 5. Watch the balloon expand!

What just happened?

As the yeast and sugar react, it releases carbon dioxide. The gas fills the bottle and then fills the balloon as more gas is created.

Balloon Blowing

Duration of Experiment: 15 mins.

Difficulty Level:

What you need:

Antacid

Balloon

Cooking vinegar

Plastic bottle

What to do:

Step 1. Put a little vinegar in the bottle.
Step 2. Drop an antacid tablet and immediately put the balloon over the mouth of the bottle.
Step 3. Wait for about 10 minutes for the balloon to expand.

What just happened?

The vinegar and antacid react to form carbon dioxide. This carbon dioxide fills the space in the balloon, making it expand.

Smoke Rings

Duration of Experiment: 15 mins.

Difficulty Level:

What you need:

Scissors

Tape

Cling wrap

Talcum powder

Shoe box

What to do:

Step 1. Cut off one side of the shoe box and replace it with cling wrap.

Step 2. Cut a hole on the opposite side of the box.

Step 3. Put some talcum powder in the box.

Step 4. Tap on the plastic covered side and watch the rings form!

SCIENCE AROUND US

Vortices

The scientific name for your 'smoke rings' is vortices. Vortices occur when liquids or gases spin or flow in a circular manner. Vortices usually occur in water but they sometimes occur in air as well.

What just happened?

The powder takes the form of doughnut-like shapes on its way out. Because the powder moves slower than the air, it maintains its shape.

Smoggy

Duration of Experiment: 10 mins.

Difficulty Level:

What you need:

Matchsticks

Ice cubes

Jar

Paper

Aluminium foil

Water

What to do:

Step 1. Make a 'lid' for the jar with the aluminium foil.

Step 2. Take it off and wet the insides of the jar with the water.

Step 3. Put a few scraps of burning paper in the jar.

Step 4. Immediately seal the jar with aluminium foil and put the ice cubes on top. The air inside the jar will soon look smoggy.

What just happened?

The burning paper caused some of the water in the jar to evaporate. This water vapour then rose and condensed back into drops of water when it came into contact with the foil.

Air Scales

Duration of Experiment: 10 mins.

Difficulty Level:

What you need:

Matchsticks

Stick

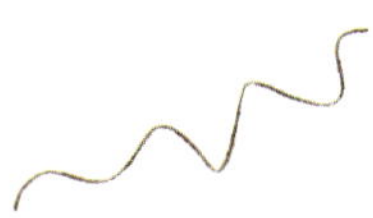
String

Candle

Paper bags

What to do:

Step 1. Hang both the bags on the two ends of the stick. The open end of the bag should be facing down.

Step 2. Tie a string to the centre of the stick so that it balances horizontally.

Step 3. Hold the lighted candle below one of the bags.

Step 4. Watch as it slowly begins to rise.

What just happened?

The warm air rises, lifting the paper bag along with it.

Candle at the Door

Duration of Experiment: 10 mins.

Difficulty Level:

What you need:

Matchsticks

Candle

What to do:

Step 1. Go into a warm room and leave the door partially open.

Step 2. Light the candle and hold it at the top of the door.

Step 3. You will be able to see the flame 'bend' towards the outside.

Step 4. Hold the candle at the bottom of the door.

Step 5. The flame bends 'inwards' this time.

What just happened? The flame bends 'outward' at the top of the door because warm air rises and escapes out of the room.

Stab the Potato

Duration of Experiment: 5 mins.

Difficulty Level:

What you need:

Straw

Potato

What to do:

Step 1. Try to stab the potato using the straw. It won't be easy.

Step 2. Now, place your thumb over the top of the straw and try to stab the potato.

Step 3. It's much easier to stab the potato this time!

What just happened?

Covering the top of the straw with your thumb traps the air inside, forcing it to compress as you stab the straw through the potato skin. This makes the straw strong enough to pierce the potato.

Sprayer

Duration of Experiment: 15 mins.

Difficulty Level:

What you need:

Straws

Glass

Water

What to do:

Step 1. Put a straw in half a glass of water.

Step 2. Hold the other straw at a 90° angle to this one. Blow hard.

Step 3. You will notice the water level rising in the straw.

Step 4. Blow long enough, and it will soon look like a hairspray.

What just happened?

When you blow on top of the straw, it lowers the pressure, making the water rise in the straw.

Dry Newspaper

Duration of Experiment: 10 mins.

Difficulty Level:

What you need:

Bowl

Glass

Newspaper

What to do:

Step 1. Stuff a sheet of newspaper into the glass. Make sure it is packed tight but do not let it come all the way up to the rim of the glass.

Step 2. Turn the glass over and submerge it straight into the bowl. Do not tilt the glass.

Step 3. Hold it there for 10 seconds and take it out. Your newspaper sheet is as dry as the Sahara desert!

What just happened?

There is a layer of air between the newspaper and the water. This layer exerts pressure on the water and does not let it touch the newspaper.

Deflating Balloons

Duration of Experiment: 10 mins.

Difficulty Level:

What you need:

2 balloons

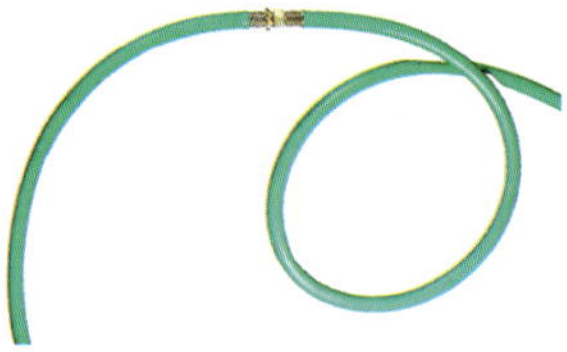

Rubber hose

What to do:

Step 1. Blow up one of the balloons to a normal size. Ask someone to hold it.

Step 2. Blow the other balloon to half that size.

Step 3. Put both the balloons on opposite ends of the tube. Don't leave the neck of the balloons.

Step 4. Once they are comfortably fitted, leave both at the same time and see what happens. Contrary to what you expect, the small balloon deflates completely.

What just happened?

Though you'd expect the big balloon to push air into the smaller one, the opposite happens. Air always moves from high pressure to low pressure. The pressure in the smaller balloon is higher than the bigger one, and so the air moves from the small balloon to the big one.

Rocket Pinwheel

What you need:

Tape

Pin

Balloon

Pencil (with an eraser on the end)

Bendy straw

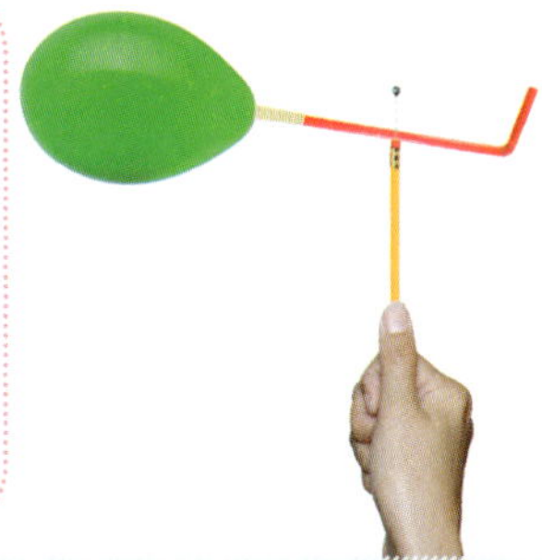

What to do:

Step 1. Tape the balloon to the long end of the bendy straw.

Step 2. Attach the straw from the middle to the bottom of the pencil with the pin.

Step 3. Blow the balloon through one end of the straw.

Step 4. Release it and watch your pinwheel spin!

What just happened?

When the air is pushed out from the balloon, it travels through the straw and comes out from the other end of the straw. It pushes against the air that is already there. This powers the pinwheel.

Balloon Surfer

Duration of Experiment: 20 mins.

Difficulty Level:

What you need:

Bottle cap

Nail

Balloon

Tub

Water

What to do:

Step 1. Make a hole in the bottle cap with the nail.

Step 2. Blow the balloon. Do not tie it.

Step 3. Stretch it over the cap, still holding it from the neck.

Step 4. Place it in a tub of water and watch it go!

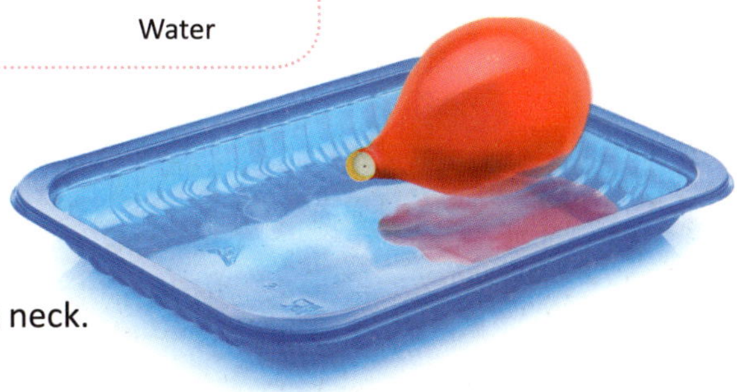

What just happened?

When the balloon is left free, the air rushes out through the small hole in the cap. This propels the entire device forward.

Balloon in a Bottle

Duration of Experiment: 10 mins.

Difficulty Level:

What you need:

Pin

Balloon

Plastic bottle

What to do:

Step 1. Put the balloon in the bottle.
Cover the mouth of the bottle with the balloon.

Step 2. Try to blow into the balloon. You will not be able to expand the balloon.

Step 3. Now poke a hole at the bottom of the bottle and try again. You can now blow the balloon.

What just happened?

Without the hole, the air inside the bottle is exerting pressure on the balloon, making it impossible to blow. Once you made a hole, the balloon was able to push the air from the bottle out to make space.

Air Lift

Duration of Experiment: 10 mins.

Difficulty Level:

What you need:

Book

Table

Balloon

What to do:

Step 1. Put the balloon on a table, but let the open end of it hang off the edge.

Step 2. Put the book on the balloon so that it is covering half the balloon.

Step 3. Now blow into the balloon. The balloon will be strong enough to lift the book.

What just happened?

By blowing into the balloon you are increasing the air pressure in it. This is greater than the air pressure being exerted on the book. So, it is easy to lift the book.

Floating Ball

Duration of Experiment: 10 mins.

Difficulty Level:

What you need:

Ping-pong ball

Bendy straw

What to do:

Step 1. Bend the straw into an L shape.

Step 2. Put the longer end of the straw in your mouth.

Step 3. Hold the ping-pong ball over the short end of the straw and blow. See your ping-pong ball levitate!

What just happened?

Moving air is at a lower pressure than still air. So, the ball is surrounded by a higher pressure. This ensures that the balloon remains over the straw.

Balloon Lift

Duration of Experiment: 15 mins.

Difficulty Level:

What you need:

Matchsticks

Paper

Balloon

Jar

What to do:

Step 1. Blow into the balloon in such a way that it is slightly larger than the mouth of the jar and tie it close .

Step 2. Light the piece of paper and drop it in the jar.

Step 3. Place the balloon on top of the jar with the tied end downwards.

Step 4. When the fire goes out, pick up the balloon. The jar is lifted as well!

What just happened?

The oxygen in the air gets used up by the flame. This lowers the air pressure inside the jar. The surrounding air outside, which is at a higher pressure than the air inside the jar, pushes the balloon into the jar and attaches itself with it.

Spraying Fountain

Duration of Experiment: 30 mins.

Difficulty Level:

What you need:

Food colour

Drill

2 straws

Clay

2 glass jars

Baking pan

Water

What to do:

Step 1. Drill two holes in the lid of one jar.

Step 2. Insert both straws through the holes—one should remain two inches above the lid, and the other two inches below.

Step 3. Seal them with the clay.

Step 4. Fill half the jar with water. Add yellow food colour to it.

Step 5. Fill half of another jar with water. Add blue food colour to it.

Step 6. Place the jar with blue water in the baking pan.

Step 7. Turn the jar with the straws upside down. Place the longer straw in the jar.

Step 8. Let the yellow water empty into the pan through the long straw. You will soon notice a pretty blue fountain!

What just happened?

As the yellow water from the sealed jar empties out through the straw, the air pressure inside the jar reduces. Air pressure then forces the blue water to spray out of the shorter straw.

Flame Killer

Duration of Experiment: 20 mins.

Difficulty Level:

What you need:

Matchsticks

Bowl

Candle

Cooking vinegar

Baking powder

Jar with a hole in the lid

What to do:

Step 1. Place the lit candle in the middle of the bowl.

Step 2. Fill ¼ th of the jar with baking powder. Pour vinegar till the jar is halfway full. Immediately, shake the jar carefully. The mixture will start fizzing and popping. Now, point the hole of the jar towards the candle flame. The flame goes off.

What just happened? A flame needs oxygen to continue burning. The baking powder and vinegar reacted to form carbon dioxide. Carbon dioxide does not allow the flame to continue burning.

Potato Cannon

Duration of Experiment: 10 mins.

Difficulty Level:

What you need:

Pencil

Potato slices

Cardboard tube

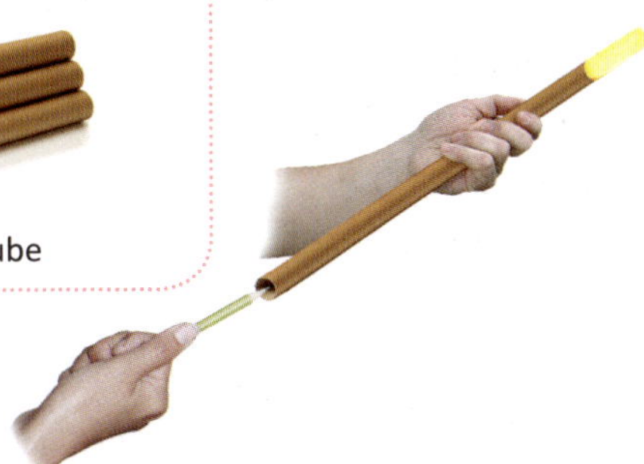

What to do:

Step 1. Push one end of the tube through a slice of potato. The slice should stick to the tube, completely covering it.

Step 2. Do the same with the other end.

Step 3. Use the pencil to swiftly push the potato from one side of the tube.

Step 4. The potato at the other end will fly out.

What just happened?

You compressed the air in the tube by pushing the potato in from one end. You just made an elementary air gun!

The Hungry Bottle

Duration of Experiment: 10 mins.

Difficulty Level:

What you need:

Matchsticks

Banana

Glass bottle

Paper

Oil

What to do:

Step 1. Tear the paper into small pieces. Dip the pieces in oil.

Step 2. Put these in the bottle.

Step 3. Drop a lit matchstick in the bottle.

Step 4. Quickly place the banana over the bottle with the fruit in the bottle, and the peels at the sides.

Step 5. Watch the bottle eat the banana hungrily!

What just happened?

The burning paper used up all the oxygen in the bottle. This decreased pressure in the bottle. The air pressure outside was sufficient to push the banana into the bottle.

Glass of Air

Duration of Experiment: 10 mins.

Difficulty Level:

What you need:

Glasses

Tray

Water

What to do:

Step 1. Pour some water into the tray.

Step 2. Lay the glass in the tray so that it fills with water completely.

Step 3. Raise the glass upside down. Make sure the mouth remains underwater.

Step 4. The water remains in the glass!

What just happened?

The air around exerts pressure on the water, which forces the water into the glass.

Sticky Plate

Duration of Experiment: 5 mins.

Difficulty Level:

What you need:

Matchsticks

Small plate

Jar

Paper

Paper towel

Water

What to do:

Step 1. Wet the paper towel and fold it into a square or rectangle that is slightly wider than the mouth of the jar.

Step 2. Lay it flat in the centre of the plate.

Step 3. Tear a piece of paper, light it and drop it in the jar.

Step 4. Turn the jar over onto the wet paper towel and hold it in place till the paper stops burning. Lift the jar. The plate gets lifted too!

What just happened?

When you drop the burning paper in the jar, the air pressure inside increases. Once this air cools, the pressure decreases. This creates a vacuum.

Stuck Glasses

Duration of Experiment: 10 mins.

Difficulty Level:

What you need:

Matchsticks

Rubber ring

Glasses

Paper

What to do:

Step 1. Wet the ring and lay it on top of one of the glasses.

Step 2. Drop a bit of burning paper into the glass and immediately put the other glass upside down over this glass.

Step 3. When the paper stops burning, you will notice that they are both stuck to each other.

What just happened?

Burning the paper causes lower air pressure inside the two glasses. The greater pressure of the air outside causes the glasses to remain tightly stuck to each other.

Marshy Mellows

Duration of Experiment: 20 mins.

Difficulty Level:

What you need:

Straw

Clay

Nail

Glass jar with metal lid

Mirror

Marshmallows

Hammer

What to do:

Step 1. Use the nail to make a hole in the middle of the metal lid.

Step 2. Put the straw through the hole in the lid and seal it with the clay.

Step 3. Drop the marshmallows in the jar and shut the lid.

Step 4. Hold the mirror such that you can see the marshmallows.

Step 5. Blow air into the jar and watch them shrink!

SCIENCE AROUND US

How are marshmallows made?

Marshmallows are mostly sugar and water wrapped around a bunch of air bubbles. This is why they are so soft and spongy. If you were to microwave a marshmallow, the air pockets would expand and the marshmallow would grow to four times its size! Try it out!

What just happened?

Marshmallows have a lot of air pockets. They are not a solid object. By blowing air into the jar, you increased the pressure within the jar, crushing the air pockets and making the marshmallow reduce in size.

Balloon Rocket

Duration of Experiment: 20 mins.

Difficulty Level:

What you need:

Tape

Straw

String

Balloon

Stool

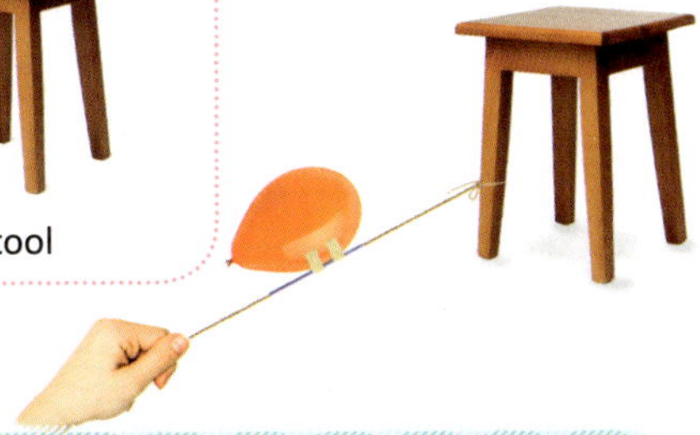

What to do:

Step 1. Tie one end of the string to a chair or stool.

Step 2. Put the other end of the string through the straw.

Step 3. Pull the string tight.

Step 4. Blow up the balloon, pinch the end and tape the balloon to the straw.

Step 5. Let go and watch the rocket fly!

What just happened?

The air rushes out of the balloon, making it move forward. This sort of force is called 'thrust'.

Windy Balloons

Duration of Experiment: 10 mins.

Difficulty Level:

What you need:

Balloon

Hair dryer

What to do:

Step 1. Blow up the balloon.

Step 2. Turn on the hair dryer.

Step 3. Point the hair dryer upward and put the balloon on top of it.

Step 4. See if you can guide the movement of the balloon with the hair dryer.

What just happened?

The moving air exerts lower pressure than the still air around the balloon. Hence, the balloon gets pulled towards the hairdryer instead of getting blown away.

Attractive Apples

Duration of Experiment: 10 mins.

Difficulty Level:

What you need:

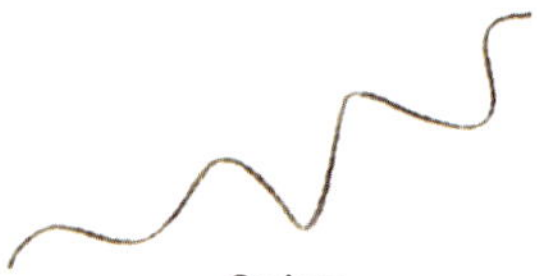
String

Apples

What to do:

Step 1. Tie the string to the stems of two apples.

Step 2. Hang the two strings to a clothesline or curtain rod, two inches away from each other.

Step 3. Blow between the apples. They get attracted to each other!

What just happened?

When you blow in the middle of the two apples, the air pressure between them becomes lower. This makes the still air push the two apples together.

Stubborn Balls

Duration of Experiment: 5 mins.

Difficulty Level:

What you need:

Paper

Plastic bottle

What to do:

Step 1. Lay the bottle horizontally.

Step 2. Make a small paper ball, about half the size of the mouth of the bottle.

Step 3. Place the ball at the mouth of the bottle and try to blow it in.

Step 4. No matter how hard you blow, the ball comes shooting back out at you!

What just happened?

Moving air has lower pressure than still air. When you were blowing on the ball, the air is deflected around the sides of the bottle. This makes the paper ball come shooting out every time you blow on it.

Collapsing Cans

Duration of Experiment: 20 mins.

Difficulty Level:

What you need:

Can

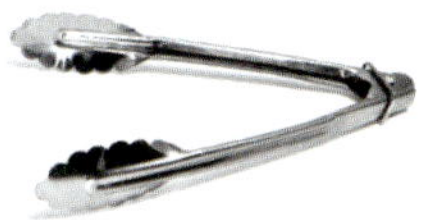
Tongs

Saucepan

Water

What to do:

Step 1. Fill the saucepan with cold water.

Step 2. Put 1 tbsp of water into the empty can.

Step 3. Heat the can on the kitchen stove to boil the water.

Step 4. Invert the can in the saucepan with the tongs. It will collapse almost instantaneously.

What just happened?

The vapour from the boiling water pushed air out of the can. When it was suddenly cooled by inverting it in the water, the vapour condensed, forming a vacuum. The low pressure in the can allowed the pressure outside to crush it.

Thirsty Candle

Duration of Experiment: 20 mins.

Difficulty Level:

What you need:

Matchsticks

Food colour

Bowl

Candle

Glass

Water

What to do:

Step 1. Fill very little water in the bowl. Add 2 drops of food colouring to it.

Step 2. Place the candle in the middle of the bowl and light it.

Step 3. Turn the glass over and place it over the candle.

Step 4. Watch as the coloured water gets sucked into the glass.

What just happened?

When the candle burns inside the glass, the air inside the glass expands, creating higher air pressure. The high pressure air inside tries to move towards the lower air pressure outside, creating space for the water to get 'sucked' in.

Can Spin

Duration of Experiment: 30 mins.

Difficulty Level:

What you need:

Nail Candle String Matchsticks Bucket Can Water

What to do:

Step 1. Make a hole in the middle of one side of the unopened can. Drain all of the soft drink from the can. Turn the nail to the right until it lies flat against the can.

Step 2. Make a second hole on the opposite side of the can. Again, turn the nail to the right until it lies flat against the can to create a hole facing the right.

Step 3. Fill the bucket with water.

Step 4. Push the can into the bucket and fill it up to a quarter.

Step 5. Tie the string to the top of the can so that it hangs.

Step 6. Light the candle.

Step 7. Hold the can over the candle till the water starts boiling and the can spins.

SCIENCE AROUND US

How do steam engines work?

A steam engine is a type of machine that gets energy from steam. Energy from a steam engine comes from the heat of the boiler, which works much like a huge pot filled with water. This is just a more elaborate form of your spinning can!

What just happened?

When water boils, it turns into steam. The steam inside the can builds up pressure and starts to push out through the holes in the side of the can. This marks the can spin.

Build a Hovercraft

Duration of Experiment: 40 mins.

Difficulty Level:

What you need:

Bottle cap

Balloon

CD

Glue

Drill

What to do:

Step 1. Make a hole in the bottle cap using the drill.

Step 2. Glue the cap to the centre of the CD.

Step 3. Seal it so that air cannot escape.

Step 4. Fit the neck of the blown balloon over the cap.

Step 5. Put your hovercraft on a smooth surface and see it effortlessly glides across.

What just happened?

The air flow created by the balloon causes a cushion of moving air between the disc and the surface. This lifts the CD and reduces the friction.

Straw-struck

Duration of Experiment: 5 mins.

Difficulty Level:

What you need:

2 straws

Soft drink

Glass

What to do:

Step 1. Fill the glass with the soft drink.

Step 2. Put one end of both the straws in your mouth.

Step 3. Put the other end of one straw in the drink and leave the second straw in the air and try to drink. It's impossible because all you get is a mouthful of air!

What just happened?

Sucking the straws creates a lower air pressure in your mouth. This lower pressure has to be filled by something. Since air is lighter than liquid, air rushes in through the second straw.

OTHER TITLES IN THIS SERIES

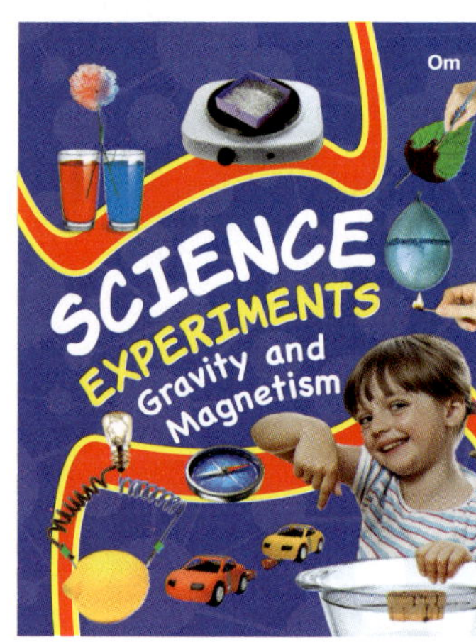